TRY DRYING IT!

Case studies in the dissemination of tray drying technology

Barrie Axtell and Alex Bush

AF192565

Practical
ACTION
PUBLISHING

Intermediate Technology Publications 1991

Practical Action Publishing Ltd
25 Albert Street, Rugby,
Warwickshire, CV21 2SD, UK
www.practicalactionpublishing.com

A catalogue record for this book is available from the British Library & Library of Congress

ISBN 978-1-85339-039-5 Paperback
ISBN 978-1-78044-368-3 Digital book

Citation: Axtell, B. (1991) *Try Drying It!: Case studies in the dissemination of tray-drying technology*, Rugby, UK: Practical Action Publishing
https://doi.org/10.3362/9781780443683

Since 1974, Practical Action Publishing has published and disseminated books and information in support of international development work throughout the world. All print editions are produced and distributed via ethical and sustainable print on demand global facilities.

Practical Action Publishing is a trading name of Practical Action Publishing Ltd (Company Reg. No. 01159018 | VAT 880 9924 76). All profits are covenanted back to its parent group, Practical Action (Charity Reg. No. 247257).

The manufacturer's authorised representative in the EU for product safety is Lightning Source France, 1 Av. Johannes Gutenberg, 78310 Maurepas, France.
compliance@lightningsource.fr

CONTENTS

LIST OF ILLUSTRATIONS

ACKNOWLEDGEMENTS

It is impossible to thank individually all those who have contributed to the writing of this book. Certain people, however, should be publicly acknowledged, especially Gonzalo La Cruz and Saleha Begum for their help in putting the book together.

In Peru, Walter Rios, Francisco Ortiz, Julian Dolorien and the staff of Yerfil.

In Colombia, Roberto Moreno Ardila and his staff.

In Guatemala, Dr Mario Molina, the staff and members of both INCAP and Cuatro Pinos.

In Bangladesh, Karl and Ilene Bergen, Judy Edmister, all at the Mennonite Central Committee and the members of the Surjosnato co-op.

Currency conversions used:

US$ 1.00 =   Q 2.50 (Guatemala)
             I 37.00 (Peru)
             Pesos 245 (Colombia)
             Tk 35 (Bangladesh)

FOREWORD

Small-scale mechanical drying is a technology that has created a great deal of interest, and at the Intermediate Technology Development Group (ITDG) we regularly receive requests for information and assistance in this area. There is undoubtedly a gap in the equipment market between the many solar (and assisted solar) drier designs that have been promoted by the appropriate technology movement and the larger industrial equipment. In this book we recount some of our experiences with small-scale mechanical drying and set out some arguments as to why this is an area that deserves more attention.

We begin with a short history by Barrie Axtell, ITDG's Food Processing Programme Manager, of his involvement with this technology, which explains how the original drier design was developed and the genesis of some of the projects with which ITDG has been working. This is followed by case studies of projects in Guatemala, Peru, Colombia and Bangladesh. Each of these studies illustrates different points about the technology and the ways in which it can be used. In the next section, we describe recent developments and the plans for ITDG's tray drier programme. A short chapter following the case studies attempts to draw together the experience we have gained in the four countries and to relate it to ITDG's overall objective of assisting resource-poor producers in developing countries.

There follows a discussion of the economics of small-scale drying in which Saleha Begum defines the difference between drying for preservation and drying for product refinement and argues that only the production of relatively high-value products will be able to support investment in mechanized drying on a small scale.

We then look at technical options in drying equipment, briefly describing the advantages and disadvantages of each type of drier with particular reference to capacity and capital cost. Finally, the Appendix contains a very simplified introduction to the theory of drying.

This represents what might be termed a first generation of information on the application of small-scale drying technologies. It does not allow the reader to go any further in terms of establishing plant. A range of more detailed technical books has thus been prepared, details of which can be obtained from ITDG.

INTRODUCTION

by Barrie Axtell
Food Processing Programme Manager
ITDG

This book is essentially a history of the development and dissemination of a small-scale industrial tray-drying system as part of the Intermediate Technology Development Group's programme of assistance to developing countries. We also hope to show how a technology develops and adapts when evolved under real-life field conditions.

Drying is the oldest method of food preservation known and remains the most common, particularly in developing countries. When people did not have a wide range of packaging materials to protect the food that they dried, products had to be compatible with local climates. Nowadays people all over the world have been influenced by new food habits; markets far distant from the areas of production have developed, and non-traditional crops are being cultivated.

To keep pace with these changes, a range of new technologies has had to be developed. Work has been carried out in two broad areas: comparatively low-cost solar driers and higher technology, large-scale drying systems including spray and freeze driers. Much of the solar drying research and development has been aimed at Third World countries, but simple solar driers suffer from the disadvantage of not operating at night or during inclement weather. To date there seems to have been less emphasis on developing small-scale versions of the higher technology types of drier capable of producing good quality products economically, in a routine production environment, and appropriate for use in developing countries.

When working in the field there is rarely sufficient time to develop fully researched solutions to a problem. Indeed the problem is usually ill-defined to begin with and I have found that the old adage 'necessity is the mother of invention' applies. The objective is to come up with an answer to a real problem, fully realizing that the system developed will later be modified by other factors which may well not have been identified or even recognized at the conceptual stage. Aspects such as fuel costs, social working patterns, and increasing demand gradually modify the production system to suit local conditions.

1

Another aspect in the development of a production technology perhaps not sufficiently appreciated is that, while theoretical feasibility studies can obviously give broad indications on viability, they rarely provide the full answer. When a product is manufactured for sale its marketability, in terms of quality and price, is central to success. What the trade demands is a costed sample of the product to assess, prior to making any commitment to purchase. But, in order to provide samples, it is necessary to produce. Through actual manufacture, data becomes available on raw material costs, fuel costs, labour costs etc., allowing a far better estimate of real production costs.

The original batch drier in St Vincent

Since, at this early stage, there is always a high element of risk, small production 'starter kits' are useful. The small-scale drier technology is a good example of this concept of a starter kit. Its capacity is sufficient to provide sound production and market data without too great a capital outlay, and often to form the basis of a viable business. As markets develop, the size of the drying system can be increased in stages with comparative ease.

The story of the development of the tray drier started in 1974. Product chemists for the government of St Vincent, with whom I was working under a British government bilateral assistance programme, noted that West Indian sorrel *(Hibiscus sabdarifens)* flowers were extremely popular for making traditional Christmas drinks. The plant flowers only in December and January and for the rest of the year sorrel was simply not available. A quantity of sorrel was dried and stored away until the summer when the laboratory's manufacturing wing, the Agrolab, marketed a few hundred bottles of sorrel syrup. The results were astonishing: the product soon sold out and the local stores were hammering on the doors demanding more.

The following year, the Agrolab purchased as much sorrel as could be dried in the pilot small-scale drier available and then both sorrel syrup and two ounce packs of dry sorrel were marketed for people to make up at home. The latter, with its lower packaging cost, proved to be the most popular and within a year had resulted in substantial orders from Canada and Jamaica.

The St Vincent Ministry of Agriculture decided to promote the growing of sorrel as a cash crop, particularly in the low rainfall areas of the island's leeward coast. They found that, not only was it an ideal crop for dry areas, but it also provided shade that allowed intercropping with other plants such as tomato.

The problem then arose of quickly increasing drying capacity to meet the next harvest season, with the added pressure that the Ministry had given farmers a guarantee to purchase all sound sorrel at a fixed price. A simple plywood cabinet was constructed with angle-iron runners to accommodate twelve trays. This was then, by means of ducting, connected to the diesel-fired heater/blower of an existing cocoa drier. Working in shifts day and night and pre-drying in the sun on the cocoa-drying floors, it proved possible to process all the sorrel brought to the laboratory.

In 1978, again under a British technical assistance programme, I transferred to the beautiful Central American country of Guatemala, and was based in the Instituto de Nutricion de Centro America y Panama. INCAP was approached for assistance by a local agricultural co-operative, Cuatro Pinos, of Santiago Sacatapequez. Like many other villages of the highlands, Santiago had been badly hit by the 1976 earthquake which in a few minutes killed some twenty thousand Guatemalans. In Santiago, over 90 per cent of the houses were destroyed and a Swiss church-funded agency, the Grupo Suizo, started to provide assistance, initially in self-build house construction. This phase soon led to recognition of other problems, particularly income generation, and by 1978 the agricultural co-operative, Cuatro Pinos, had been formed.

Santiago, like many other Guatemalan villages, is characterized by the high percentage of farmers with very small landholdings. These are usually too small to supply the family needs for the traditional staples, beans and maize. The Grupo Suizo decided to investigate the potential for growing higher-value vegetables to improve income while at the same time making sure that the traditional crops, which are of considerable cultural importance, were retained. Trials were carried out with a range of temperate vegetables. Cabbage proved just too popular with farmers and a glut resulted.

3

At this point Dr Mario Molina of INCAP and I were asked to look at possibilities for processing this excess cabbage. Two lines were examined: the production of sauerkraut — for which a very limited local market existed — and dehydration. Through the work on dehydration it was discovered that several large dried-soup manufacturers, including Nestlé, were importing dried vegetables from Europe and the USA to make packet soups for the Central American market.

The range of ingredients being imported was examined and it was noticed that each packet of soup contained a pinch of dried parsley. The total demand for parsley was 14 000 kg per annum at a price of US\$8/kg, which was considerably more interesting than that for cabbage, for example, with a demand of 1500 kg at \$4.50/kg. The co-op explored the possibility of parsley production and found that it grew well in Santiago. Equally importantly, they realized that it was a highly labour-intensive crop to harvest and thus more suitable for small (market garden size) units than larger farms. This probably accounted for the high ex-Europe price.

It was agreed that the production of certain dehydrated vegetables, including parsley, warranted further investigation. More accurate production costings were needed and the co-op's agronomists were asked to investigate the growing of parsley on a larger scale.

A simple plywood double-cabinet drier was built and set up in the INCAP pilot plant. Each cabinet contained nine trays, and heated air was supplied to the drying chamber by a thermostatically controlled, indirect, Benson-Jetaire heater/blower unit which was imported from the UK. This heater, with a maximum output of 60kW and air flow of 1700 cubic feet per minute (cfm), was able to maintain temperatures of up to $80^{\circ}C$ in the drying chamber and, being indirect (that is, with a built-in heat exchanger), did not allow the products of combustion to pass over the product.

Two main technical problems emerged that required a solution. First, the buyers all had strict standards regarding microbiological quality. These proved to be difficult to meet and it took about eighteen months before a suitable washing and sanitizing system was finally developed. The second problem concerned the specification on the percentage of small stalks in the dry parsley. Eventually a modified winnowing machine enabled INCAP to meet the buyers' requirements in this respect.

The pilot plant drier ran for almost a year, gradually evolving from a test drier to a production unit. This enabled the co-op to use the facility to train its members in all aspects of the process including marketing the final product. Gradually the co-op took over responsibility for supplying

The prototype drier at INCAP, Guatemala

all labour, raw materials, fuel etc. These costs were largely being met by income from sales.

Certain characteristics began to show up in the design of the prototype batch drier. It was soon realized that the lower trays, being presented to the hot air first, dried soonest. They could therefore be removed, the trays above could be lowered, and new trays of fresh material could be inserted in the space left vacant. In this way outputs were increased to meet the demand for the product and the greater inflow of raw material. Eventually the system had to go on to a shift basis — often working twenty-four hours a day all week. The continual removal, lowering and replacing of trays by hand soon became an inconvenience and everyone's mind turned to a means of automatically moving trays without resorting to sophisticated systems.

At this point the concept of a semi-continuous tray drier (SCTD) mechanism arose. A set of metal fingers, operated by a lever, allowed all the trays except the bottom one to be lifted by a few inches, so making it possible to remove the bottom tray of dry material. On release of the lever, the remainder of the stack of trays could be lowered and a tray of fresh material loaded into the empty space created at the top of the cabinet.

By this time the Cuatro Pinos co-op had decided to establish its own, somewhat larger, drying facility at Santiago. Funding was obtained from the EC and a semi-continuous double cabinet linked to a somewhat larger (400 000 BTU) Benson-Jetaire heater/blower was installed. The Santiago drying plant is described in fuller detail in the relevant section of this book.

The technology was transferred to Peru largely through the interest of an ex-student of INCAP, Walter Rios. He felt that the drier had applications in his country and had identified a small entrepreneur with an interest in drying. A visit was made to Peru and an agreement made with the owner of the plant whereby ITDG would fund a heater/blower, and the collaborator would fund the construction of an SCTD cabinet. It was also agreed that other interested parties would be able to examine the drier when required.

At this stage only very rough sketches of the INCAP tray-moving mechanism were available. These were taken to a small workshop in Lima, Dolorien Engineering, and after a number of traumatic experiences the first cabinet was finished in a week. The construction of this cabinet provided ITDG with enough data to prepare a set of working drawings for cabinet construction. The drier was set up in Lima and a little later moved to the main herb production area at Tarma in the Sierra.

In 1985 ITDG had funded a three-day workshop on drying, hosted by the International Potato Centre in Lima. Participants included representatives of NGOs, universities and entrepreneurs. ITDG funded the visit of Dr Mario Molina, who had been instrumental in the initial implementation of the technology in Guatemala, to act as principal speaker. This workshop publicized the technology widely in Peru and over the next two or three years the tray drier rapidly took on a life of its own. By 1988, with the help of Walter Rios, who was by now working as an ITDG project officer, 25 units were operational in Peru and a locally designed and fabricated heater/blower was being marketed by Dolorien Engineering.

The system also adapted in interesting ways to local needs and conditions. In almost all cases the users rejected the semi-continuous cabinet design, largely because of its cost. Kerosene costs were fairly low in Peru and it was perceived as more important to reduce initial capital investment costs than to save fuel by the use of the more energy-efficient semi-continuous cabinet. It was discovered that Volkswagen vans were being imported in large crates that could be converted into ideal batch drying cabinets—and they cost only $300. The norm in Peru has thus become a wooden drying chamber linked to a locally made indirect heater/blower working on a batch basis. Dolorien has now acquired a name for his knowledge of drying and has even produced a drier for fixing the dyes on printed fabrics.

A major test of ITDG's drawings of the drier cabinet occurred in 1985. A letter was received from a company wishing to dry a range of herbal teas in Bogota, Colombia. They simply could not justify the purchase of an off-the-shelf drier; the lowest quote obtained was $40 000. A set of drawings and information was sent in response to this enquiry and, a couple of months later, ITDG received a letter and photographs saying that the unit was up and running. The drawings really did work!

ITDG also agreed to collaborate with the University of Ambato in Ecuador to establish a drier with a local fruit-growers co-op. A heater/blower was shipped out but the installation was considerably delayed by a serious earthquake. The local engineer who was to construct the cabinet did not fully understand the drawings — clearly more work was required in that area. A visit by the engineer to Peru to see the drier in operation soon solved all problems. ITDG were very happy to see problems being resolved by interchange between Peru and Ecuador and to hear that the drier has been constructed.

At about the same time Bangladesh was appearing as a possible site for a drier. For a number of years, ITDG had been involved with the Mennonite Central Committee (MCC) which had been helping a group of landless women set up a small industry to dry and market desiccated coconut, mainly to commercial users such as bakeries. The initial project was designed around drying carried out in simple, locally built Brace-type solar driers. Sales gradually increased and in 1986 ITDG received a request for technical advice and support. It appeared that the buyers of the desiccated coconut in Bangladesh were insisting that the MCC group should supply dry coconut all year. If this could not be done, they would be forced to import in order to have continuity of supply. The reliance on solar driers meant, however, that the group was unable to produce in the monsoon season. Some form of artificial drier was thus needed. A wood-fired tray drier was finally set up and has largely solved the group's problems.

An interesting and somewhat surprising request for the tray-drying technology came from Ireland, not a country that ITDG would normally expect to work in. Here a small, low-cost but controllable drier was needed to explore the agronomic problems, production costs, and market potential for organically grown herbs. The SCTD provided a suitable, low-risk test kit that enabled the Shannon Development Authority to carry out the needed development work.

The following chapters outline a continuing story in development from which those involved in such work may find some useful lessons. For the more general, interested reader it offers an insight into the problems and challenges faced by those at the sharp end of development work. One thing seems certain: there is still more work to do in this area and, at the time of writing, ITDG was seeing how interested collaborators in three new countries can be assisted.

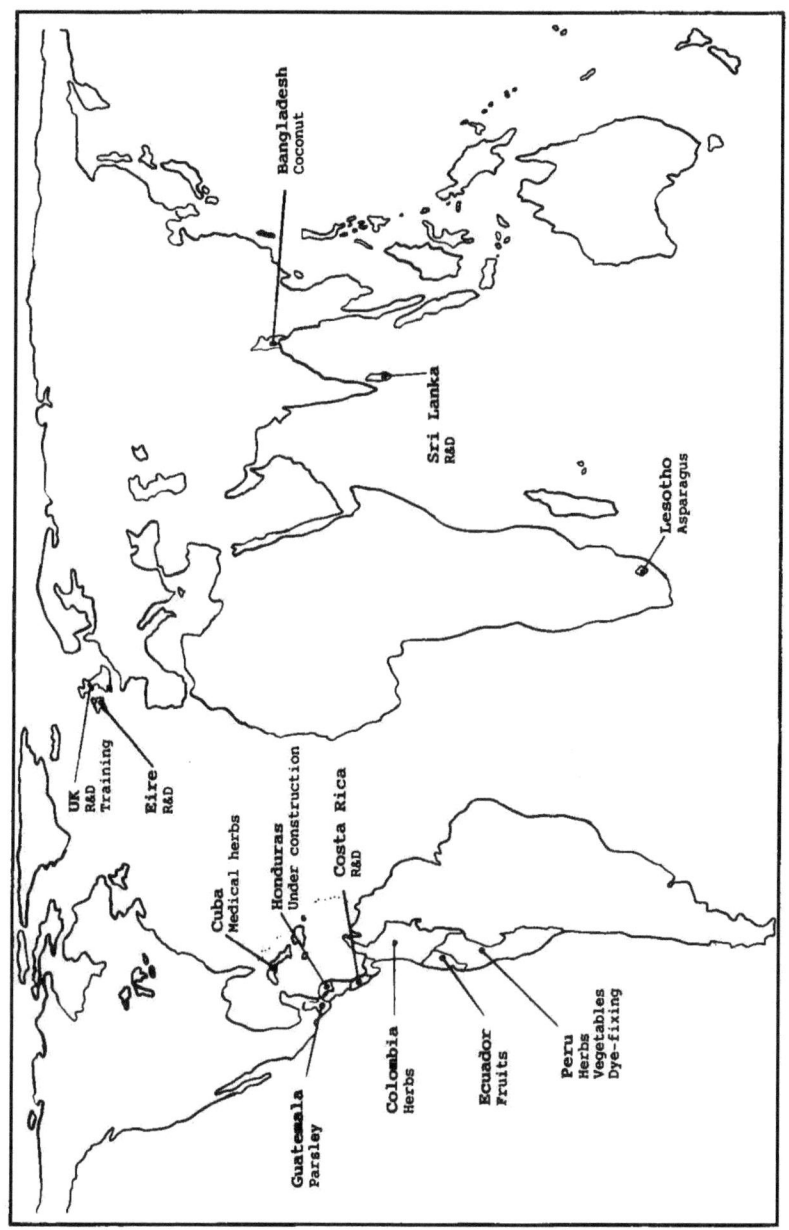

The tray drier: worldwide use for research and production

1. CASE STUDIES

GUATEMALA

The story of the ITDG tray drier began in Guatemala. We should perhaps refer to it as the INCAP drier since that is where most of the initial development work was done. Barrie Axtell's introduction tells the story of those early years; this chapter discusses the running of the drying plant at the Cuatro Pinos co-operative since then.

The co-operative is situated in Santiago Sacatapequez, twenty-five kilometres from Guatemala City, and has 1400 members and 490 hectares of land. Its main business is exporting garden produce (green beans, cauliflower, broccoli, mange-tout etc). Thirty or so members grow parsley on about seven hectares of land and the co-operative has around 40 per cent of the local market for the dried product. The income from the drying business accounts for only 0.5 per cent of the total income of the co-operative. Production of dried parsley is about five tonnes per year.

The dried product is sold to a local manufacturer of dried soups. Their quality control requires a good colour in the final product, flavour being rather less important than in the other case studies.

Technology

The co-operative's drying plant is the largest and most technically sophisticated of the operations described in this book. It consists of ten small batch cabinets connected to two direct gas-fired heaters, a fully mechanized washing machine, and two modified winnowing machines for separating leaves and stalks. This, together with office space for the supervisor, is all situated in a large, well-constructed concrete building with tiled floors.

Cuatro Pinos had great difficulties early on in meeting the microbiological specifications laid down by their customers. Today the plant has a strict system of control. Parsley cut in the fields is immediately placed in plastic crates out of contact with the soil. It is then washed several times in water and solutions of calcium hypochlorite and ammonia, and finally rinsed clean. The tiled floor is regularly hosed down and the interior of the driers and all work surfaces are sprayed each day with formaldehyde solution.

The other major technical difference between this plant and the other herb-drying operations is the temperature at which drying takes place. Air

The batch drying cabinets at Cuatro Pinos

is introduced to the driers at temperatures in excess of 80°C. This has the effect of destroying an enzyme (chlorophyllase) which breaks down chlorophyll and thus the vivid green of the fresh parsley is preserved. Flavour is less important in this drying operation than in many others. Flavour is largely a function of essential oil content and high temperatures tend to drive off these oils, so these elevated temperatures may not be an option.

Cuatro Pinos originally used heaters of the indirect, diesel-fired Benson type in their plant but the resident engineer predicted substantial savings by switching to gas. Comparative tests run in March 1986 suggested that these savings would be in the order of 70 per cent. Thus direct gas heaters and a bulk gas storage tank were installed. Redesign of the fan and much reduced pressure drops resulting from the simpler configuration have led to substantially greater air flows than in, for example, the Peruvian designs.

The co-operative is marketing a product of very high quality, both in terms of microbiology and colour, and there is no reason why this should not be repeated in other plants. The higher air temperatures are more easily achieved by the direct gas heaters but should be within the reach of Benson-type devices. The microbiological control that is being exercised does not depend on sophisticated equipment but rather upon a few inexpensive chemicals and a great deal of care.

Economics

Energy is more expensive in Guatemala than in either Peru or Colombia, with gas being the cheapest source. In addition, the relative sophistication of the plant is reflected in high overheads and depreciation compared to the Peruvian plants. It is interesting to compare the economics of this plant with Yerfil in Lima. In the Guatemalan case the downstream processing is hidden; in other words we cannot evaluate the economics of the packaging process (in this case the production of dried soups). The drying process shows healthy but not spectacular returns. Profits after administration and depreciation costs in Guatemala are around 16 per cent of sales (compared with 24 per cent in the Tarma, Peru drying operation, 5 per cent in Bangladesh).

The ratio of selling price to raw material price is three to one. The farmers make approximately 20 per cent profit on the fresh parsley that is sold to the co-operative; in other words, the farmer receives 7 per cent of the final income directly as profit on the growing operation. They also receive a share of the co-operative's profit at the end of the year. The breakdown of costs is shown in Table 1.

Table 1. Cuatro Pinos costs

Item	$US	%
Farmer's inputs	1.44	25.7
Farmer's margin	0.38	6.8
Labour in plant	0.62	11.1
Fuel, chemicals, etc.	1.43	25.5
Fixed costs & dep'n	0.83	14.8
Co-op's margin	0.90	16.1
Selling price	5.60	100.0

(Unit costs per kg of final product calculated at $US1 = Q2.50, 1987 prices)

Assuming that the co-operative's margin can be counted as a benefit to the members (although some of it will go to paying salaries of the management), then 34 per cent of the final selling price finds its way back into the pockets of the co-operative's members and workers.

Organization

It is also instructive to look at the level of support that has been necessary for the co-operative to weather the difficult early years after the earthquake, and the organizational pattern it has now taken on.

When it initially decided to get involved in the export vegetable market, the co-operative made several wrong decisions as to the crops it should be growing. The present product mix has been determined largely by trial and error. Luckily for this particular co-operative, solid international funding gave it the security to live through the errors.

Today, Cuatro Pinos appears to be organized along the lines of a marketing co-operative with the central organization making the decisions

The parsley washing system used before the mechanical washer was introduced

on how much of each crop to grow and organizing seed and fertilizer supplies but leaving the individual members to work their own land.

Management of the co-operative now seems to be in the hands of the agronomist and an engineer who are employed by the members to run the day-to-day operations. They also arrange for the sale and shipping of produce, most of which goes to the USA. Involvement of the members appears to be restricted to policy decisions through general meetings.

This is in sharp contrast to the type of co-operative organization we see in Bangladesh where the members are centrally involved in the day-to-day running of the operation. How much this is simply a function of the size of the enterprise is difficult to say, but it undoubtedly plays a role.

PERU

The first Peruvian drier was built in collaboration with the Yerfil company which produces a range of herbal teabags in Lima. Initially, this company had been importing dried herbs from Argentina, Chile, Germany and elsewhere for packaging in Peru. Devaluation of the local currency, the Inti, forced Yerfil and other producers to look for domestic sources of the dried herbs.

Herbal infusions such as camomile tea are traditional drinks in Peru. In the Andean region, many people grow a small quantity of camomile at the edges of their fields and a substantial amount is traded in the local markets. It was soon found, however, that insufficient quantities of herbs were being grown in Peru and that there was no locally available technology for bulk drying.

A former student of the Guatemalan co-operative, INCAP, who was working in Peru, contacted ITDG for some technical assistance, and ITDG agreed to send a Benson heater and the plans for a cabinet. Meanwhile, the manager of Yerfil was experimenting with a variety of simple drying systems and began to buy camomile grown in the Andean region of Peru. The timing of ITDG's inputs was fortunate since an accident with the old heater burnt the factory down just before the Benson arrived.

The plans provided were for a continuous cabinet like the one developed in Guatemala, to be linked to the Benson heater. However, it seemed likely that Yerfil would be using the batch drying system, so a large cabinet, capable of holding 240 kilograms of fresh camomile, was constructed. Work on the continuous cabinet was continued at a small workshop in Lima since it was thought to have some value as a prototype in Peru.

The new cabinet and heater combination proved successful for drying the camomile in sufficient quantities but other problems remained. Yerfil were attempting to re-present this traditional product in a refined form to the middle-class market in Lima by selling it in the form of teabags. Entry into this market, however, demands high standards of hygiene from the producer and Yerfil were having great trouble meeting these standards.

The central problem appeared to lie in the fact that the camomile was being transported from Tarma in the high Andes to Lima on the coast before being processed. This was resulting in a delay of about forty hours between cutting and drying. ITDG recommended to the manager that he move the drying operation to Tarma, enabling him to establish better quality control and to deal with the producers directly. This was done and

Traditional sun-drying of camomile in the Peruvian Andes

two driers now (1989) ope
The machine in Tarma dries the locally produced camomile and has a sideline in potato drying. The drier in Lima dries a variety of herbs such as lemon verbena, lemon balm and lemon grass from different regions of Peru. All the dried herbs are packaged into a range of products in the Lima plant.

Several more driers have since been built in Peru. The Lima workshop produces heaters based on oil burners imported from the USA at a cost of about US$1100, compared to US$2200 for the Benson model. Materials being dried include various herbs, cochineal, onions, and other

The locally designed heater/blower unit in Peru

fruits and vegetables. In addition, one drier has been converted for the thermal fixing of dyed cotton cloth by replacing the trays with a set of rollers for the cloth to pass over.

Technology

The history of the drier in Peru is one of adaptation and innovation. Right at the start, the continuous cabinet was rejected in favour of the batch cabinet; a Peruvian version of the Benson heater and an entirely new application in dye-fixing have been developed. A variety of businesses have bought the equipment to dry a wide range of products.

The principal driving force behind these advances seems to have been the need to reduce the capital costs of drying equipment. ITDG in turn has learnt a lot from the experience in Peru. The drier has always been aimed at the small processor for whom capital is often the scarcest resource. The main lesson has been that the technical efficiency of the process is rarely of much significance. Saving 10 per cent on fuel is of little concern to a business where fuel accounts for perhaps 2 per cent of the final selling price. Marketing and promotion activities, for example, are likely to provide a better return to the entrepreneur's time which is often subject to heavy and conflicting demands.

Technical efficiency has only really become an issue where the drier has come close to the limits of its capacity. In general, the large batch cabinets are oversized for the local heater because the air delivery of this device is well below that of the Benson. Thus ITDG's input can perhaps be most valuable in areas such as improving the fan design in collaboration with local fabricators rather than in incremental inputs to particular plants.

As an illustration of the cost savings that have been achieved, Table 2 compares the cost of the first continuous cabinet built in Lima and the Benson heater ('imported' technology) with that of the simple batch cabinet and the Peruvian heater design ('local' technology). It should be noted that these figures do not include labour or installation costs and that these original cabinets were built with second-hand materials for trial purposes only.

Table 2. Yerfil equipment cost comparison

	Imported	Local
Cabinet	124	53
Heater	2200	1100
Total	2324	1153
	(All costs US$)	

Economics

At the plant in Tarma, five kilograms of fresh camomile are required to produce one kilogram of the dried product. This is then sold to the Lima operation in bulk for packaging as teabags. Table 3 (page 18) shows how the final selling price of 450 Intis (US$12) per kilogram is made up.

As Saleha Begum points out in Chapter 4, drying is only one component of a larger processing operation. In this case the major component of the added value is obviously provided by the packaging process. The increase in value is really quite remarkable; each Inti spent on fresh camomile becomes 2.6 Intis worth of dried camomile and, by the time it is packaged and wrapped, it is worth 18 Intis. Although some of the costs are also quite high, the total margin of 42 per cent on sales looks very healthy.

17

These figures can be compared with those for the project in Bangladesh, where the ratio of selling price to raw material cost is 1.45 and the margin around 5 per cent of sales. In terms of cost structure, Yerfil's raw material costs account for about 10 per cent of total costs, whereas the figure for the Surjosnato operation is close to 70 per cent.

Table 3. Yerfil: costs and margins

	—— Amount ——		—— % ——	
Raw material	0.68		5.56	
Tarma costs	0.69		5.72	
Tarma margin	0.39		3.17	
Price to Lima		1.76		14.45
Lima costs	5.63		46.27	
Lima margin	4.78	10.41	39.28	85.55
Final selling price		12.17		100.00

(All costs in 1987 US$ and including depreciation)

This totally different cost structure reflects the type of market into which Yerfil is selling. Lima is a sophisticated city with a well-developed consumer goods market including many types of convenience foods. Consumption patterns have moved much further towards those of Europe and America than those in Bangladesh. Thus food-processing industries can support quite high costs because high prices can be fetched for the final products.

From ITDG's point of view, the key concerns are the returns to the farmers and the workers in the project. There is a great deal of variety and flexibility in the traditional farming practices in the Andes. Our experience in the Picoy valley near Tarma bears this out. From being an almost incidental crop grown along the sides of fields, camomile has developed into a major cash crop. Fields of up to a quarter of a hectare entirely given over to camomile can be seen, and intercropping with vegetables and flowers is becoming common. Farmers are beginning to invest in the crop by applying fertilizers, pesticides and weed control. One farmer who was previously growing flowers as a cash crop claimed that he had treated camomile as a weed among the flowers but that the situation has now been reversed.

There are, then, strong indications that camomile is a very attractive crop to the poorer farmers. With average areas under camomile of three

The Peruvian batch cabinet

tongos (1 tongo = 769 square metres) per farmer, it is estimated that the gross profit after allowing for labour at local rates and other inputs is in the order of US$800 per annum. However, the key point is not the absolute level of benefits but the additional benefit gained by the farmers through substituting camomile for other crops. This proved very difficult to establish but the strong interest expressed by the farmers suggests that the substitution is profitable.

The drier's impact on employment is difficult to assess for the same reason. Directly, the drier creates only two or three jobs. However, the packaging operation in Lima is quite labour intensive — it employs ten women. There are also other employment links in terms of the manufacture and printing of the packaging materials and the work created at the fabricator's workshops. Whether these are genuinely new jobs is very difficult to determine.

A report to ITDG on the development of the Yerfil drying business suggests one further point of interest. It seems that the Tarma plant was the first agro-industrial operation to be based in that town. Yerfil's request for a bank loan and the realization that the government is prepared to provide some financial breaks to non-traditional agro-industries have raised a lot of local interest in further industrial development.

Other driers

There are many driers in Peru, operating on a variety of different products. The dye fixing unit is not strictly a drier; it fixes the dye in cotton cloth by raising its temperature to 150-200°C. It is, however, interesting as an adaptation of the batch cabinet in that it demonstrates the ingenuity of the local engineers and the type of innovation that is being made in Peru.

In this 'drier' an electric motor drives (via a chain) a series of rollers which feed the cloth into the cabinet. Inside the unit the cloth passes up and down over rollers at the top and bottom of the cabinet, remaining inside for approximately three minutes; it then passes out of the cabinet and is rewound onto a cardboard former. Cloth is delivered in 100-metre rolls and processed at about four rolls per hour. In mid-1987, the plant was operating for only two days per week but a sharp increase in activity was expected with the amount of cloth for fixing likely to rise from 1300 metres per day to 5000 metres.

Heat is supplied to the unit by a locally built burner with the thermostat removed. The burner is barely able to maintain 150°C in the cabinet—not surprising since it was only designed to reach 70°C or so. This type of heater is also designed to operate with a large flow of air across the heat exchange surfaces and is not really appropriate for this duty where temperature is critical and air flow negligible. As a result the casing of the heater is beginning to disintegrate.

The owner of the plant is extremely pleased with the unit, which he claims has led to considerable savings in energy costs compared to his previous system using a bank of infra-red lamps. Again, the 'drier' appears to be forming an important link in a relatively complicated process.

COLOMBIA

Many of the supermarkets in Bogota now display small pine boxes containing a selection of fresh herbs wrapped in cellophane. These are the product of Morenos Ltda, a family firm growing and processing herbs for sale both fresh and dried. Under the brand name *Kiska* the company also produces a range of attractively packed and presented herbal teabags.

This operation provides an interesting comparison with the Yerfil company in Peru since Morenos grow the bulk of the herbs that they dry on their own farm but contract out the packaging. They also sell substantial quantities of dried herbs in bulk. Morenos control 10 per cent of the dried herb market and about 3 per cent of the herbal teabag market in Colombia. They also have 60 per cent of the fresh herb market in Bogota.

Originally Morenos packaged imported tea for an English firm. When government legislation began to make this more difficult, they began to produce herbal teas and to build up a trade in fresh herbs. The major impetus for this development came from Roberto Moreno Ardila who acts variously as agronomist, engineer, public relations officer, financial controller and managing director of the company. He has been responsible for developing the four-hectare family farm, installing the drying equipment in Bogota and building up the marketing operation.

Technology

ITDG supplied only drawings to Colombia. Morenos went ahead in collaboration with a local equipment manufacturer and built a pair of continuous tray-drying cabinets to the original design. These are powered by a single diesel-fired heater/blower unit which was also locally built. Outlet air from the cabinets is blown through a 'batch cabinet' — simply a sealed cupboard loaded with crates of herbs. The driers are run from Monday morning to Saturday lunchtime and are the only driers mentioned in this book that are in truly continuous operation. A tray of dried herbs is removed every 25 minutes and the dried material is rubbed through a wire mesh to powder it and separate unwanted stalks etc.

Two technical problems have become apparent — quality control and capacity. Sr Moreno suspects that the market for herbal teas in Colombia is approaching saturation and he has been looking for other outlets for some time. However, the most obvious purchasers, the large food-processing companies, have very strict quality standards (as the co-operative in Guatemala discovered). Overcoming these is more a matter of scrupulous care, particularly at the cutting stage where the herbs can absorb bacteria from the soil, than of complicated chemical treatment. Simple chemical treatment during processing can do the rest.

The capacity problem is rather more complicated since there is no room for expansion at the existing premises. An ITDG evaluation visit in 1987 recommended several minor improvements in equipment and operating practice that would increase capacity. Morenos is the only company examined in this book with both the potential and the incentive to increase the throughput substantially.

Economics

After administrative and depreciation costs, the level of profit in the Morenos operation is 20 per cent of sales. The breakdown of costs per kilogram of dried product is shown in Table 4.

Table 4. Cost breakdown—Morenos Ltda

Item	Cost	%
Raw material	1.97	17.2
Drying costs	0.68	5.9
Packing costs	1.93	16.9
Sales & admin	4.27	37.4
Depreciation	0.33	2.9
Margin	2.25	19.7
Selling price	11.43	100.0

(Costs in $US per kilogram of dry product at 1987 rates)

This situation appears to be midway between the two operations in Peru and Guatemala. We do not know the selling price of bulk dried herbs and cannot therefore apportion the profits in the way they are presented in the Peruvian case. The major point to be made is that most of the profits from both the growing (some herbs are bought in from outside sources) and drying operations accrue to the owners of the company. In terms of benefits to the poorer sections of the community, the impact of this drier is limited to the creation of seven full-time jobs on the farm and two in the plant.

This case is a good indication, however, of the demands on the entrepreneur. With Sr Moreno attempting to be agronomist, engineer, accountant and salesman, he obviously has to rationalize the use of his time. With drying costs, of which fuel is only a part, accounting for only

The drying equipment at Morenos Ltda, Bogota

6 per cent of his total costs, he was right to have little interest in modifying the drier to save perhaps 10 per cent of the fuel. Moreno's real interest was in increasing the capacity of his drier with minimal capital outlay, thus spreading his administration costs more widely and increasing his margin. With marginal costs accounting for only 40 per cent of the final selling price, every extra kilo dried would realize approximately seven dollars profit.

Again, as in the case of Yerfil in Peru, a relatively sophisticated product is being manufactured for an urban market. The levels of packaging and marketing required would not be easily attained by a rural-based co-operative enterprise.

23

BANGLADESH

In rural Bangladesh, with its mostly Muslim population, women are traditionally confined to activities in and around the household. This situation does appear to be changing as a result of increasing impoverishment and landlessness. (Indeed the Bangladesh Institute of Development Studies has estimated that between 50 and 75 per cent of landless households have a female member in paid employment.) Solar driers in particular, but also the tray drier, have provided women with a source of income so they can support their families.

Chard Bla Nath is a widow living in Ramganj, Bangladesh, and is the main income earner for a household of five. There are no male earners but her fourteen-year-old daughter earns Tk200 ($6) at the local battery factory. They live in a rented house which consists of one room with a tin roof, bamboo thatch walls and mud floor. They own no land.

Chard previously earned money by weaving saris which she sold for Tk20 each, with an average production of one sari a day. She worked in a disused building but, in 1980, was evicted when it was bought by what is now the battery factory. She was forced to sell her equipment.

Before Chard lost her job the combined income of the household was Tk680 per month (around $20); afterwards they had a highly inadequate income of Tk200 per month which is not even sufficient to maintain two people above the minimum calorie requirement. (Estimates indicate that an average monthly income of Tk150 ($4.30) per person is needed to meet the minimum calorie requirement of 2273 Kcal/day.)

In 1980 Chard joined a small project run by the Mennonite Central Committee (MCC) which was producing desiccated coconut for supply to biscuit manufacturers in Dhaka. The Surjosnato (rays of sunshine) project equipped her with everything necessary to process and dry the coconuts at her own home, including a small solar drier. She is paying back the cost of the equipment and still realizes an income of about $18 each month. This project has built from small and tentative beginnings to become a well-established rural industry. Currently, it employs eighty women in coconut drying and coir processing and has an annual turnover of around $77 000. It sells the desiccated coconut to five major biscuit manufacturers in and around Dhaka.

MCC have supported the project in many ways for many years, not least with injections of capital. Financially, however, the project is now independent. Although the MCC loans are interest free, they must be (and are successfully being) repaid from income.

A solar drier at Surjosnato, Bangladesh

Initially, the business was entirely dependent on solar drying. A modified Brace-type drier was being supplied to every woman and these proved very successful. However, in the monsoon season, although solar drying was still possible in the mornings or on fine days, the average level of production suffered severely. As the project developed and the links with the biscuit manufacturers were established it was realized that, in order to keep its major customers happy, the project would have to maintain a reliable supply throughout the year.

Thus a low-cost technology was being sought for monsoon season drying. Since this was effectively a pilot project there was no previous experience to guide it. One of the food technologists attached to MCC visited England to conduct some trials with the continuous tray drier. He saw this as a possible solution to the rainy season drying problem and perhaps the only one that would be economically feasible for the project.

He left the UK with a set of plans and an undertaking from ITDG to underwrite the costs of the first drier since it was seen as an experimental unit. ITDG also agreed to send a heater unit from the UK to Bangladesh.

Construction of the cabinet was relatively straightforward but it proved impossible to import the heater unit into Bangladesh, so MCC were forced to devise their own means of heating the air. They constructed a system in which water was heated in an oil drum over a wood fire and then circulated through a tractor radiator placed in the air duct. The oil drum was later replaced by a slightly more sophisticated 'boiler'. A single small electric motor ran the water pump and the fan. This arrangement proved capable of supplying air at the desired temperatures (around 50°C) to the cabinet. The whole system including the cabinet was constructed at a cost of about $860.

Experimental work was done on the drier using fresh coconut but it was soon realized that the equipment was more likely to be used for completing the drying of batches that had been partially dried in the solar driers before rain intervened. During the 1987 monsoon season — the first in which the drier was in full operation — the producers were able to fulfil 80 per cent of the orders received and approximately four tonnes of dried coconut were produced by the tray drier in a three-month period. The drier had effectively removed the risk of losing product to spoilage in the monsoon season.

The financial success of the business has undoubtedly been mostly because of the solar driers. Economic analyses of the tray drier performed by MCC and by ITDG have indicated that, on its own, it would not be able to operate profitably. It is also undoubtedly true that the project's household-based solar driers, which do the bulk of the year's drying, are more appropriate to local traditions than a single centralized tray drier would be. By acting as an 'insurance policy' during the monsoon season, the tray drier has, however, ensured the continuing commercial viability of the project.

Organization

The fifty women employed in the project's coconut-drying operation are selected from six villages in the surrounding area (radius about one mile). Preference has been given to landless persons, widows or women responsible for supporting families, and households with only one income earner.

Each woman visits the centre once every ten days where she collects and husks an average of 300 coconuts. The coconuts are then taken back to the women's homes where the flesh is removed, grated, sulphured and dried. Sulphuring is carried out by burning a small piece of sulphur inside a polythene tent with trays of fresh coconut inside and helps to maintain the bright white colour of the final product. About thirty coconuts per day can be processed in the dry season, with a production of 3 kilograms of dried material.

Preparing coconuts, Surjosnato, Bangladesh

The day-to-day management of the business is carried out by a salaried (male) manager assisted by a (male) quality control supervisor. The manager is responsible, among other things, for organizing the purchase of coconuts and the distribution of the product. Given the cultural constraints, it would probably not be possible for a woman to fill this role.

Policy decisions and decisions on admitting new members and the distribution of profits are made by a management committee consisting of five coconut producers, two coir workers, the manager, the quality control supervisor and an MCC representative. The committee is chaired by one of the producers. In interviews, the producers expressed a strong sense of control over the project. Most felt that they could influence the management, either directly or through a committee member.

Technology
Technically there have been a few problems with the drier, which is only to be expected in an experimental unit like this. Capacities have been low and fuel costs high. As a result of an ITDG evaluation visit in March 1988 several recommendations were made for raising capacity, but the situation on the fuel side is rather complicated.

At present, heat is supplied by firewood which is increasingly scarce and expensive in Bangladesh. Owing to poor design of the firebox, the conversion efficiency of the system is very low. There is a source of fuel on-site, namely coir dust, that apparently offers an alternative to wood as an energy source. Coir dust is, however, a difficult fuel to burn efficiently on a small scale. The only feasible option would seem to be some form of briquetting process to turn the coir dust into an acceptable fuel.

There is a second option. The drier is already totally reliant upon electrical power for driving the fan and the pump. Calculations indicate that conversion to electrical heating would lead to running costs comparable with those for wood. The advantage would be that the drier could be run overnight virtually unattended, with consequent increases in capacity.

As in many rural areas of developing countries, electric power is either very unreliable or non-existent. All the driers installed to date rely on electricity to power the fan and this limits the number of suitable locations and tends to bias siting towards towns. This in turn often means that the processing moves out of the hands of the rural producers and into those of an entrepreneur, with a consequent redistribution of benefits.

It is, however, ITDG's aim to concentrate the process and its benefits as far as possible in the hands of the poor rural producer. This will not happen automatically with the development of an electricity-independent drier but such a technical improvement would be an important step forward.

In principle this development is fairly simple. The tray drier at ITDG in Rugby was run briefly using a steam system. The steam raised provided both heat and power for the process. In practice the construction of a boiler is likely to imply a large increase in the cost of the drier (in the order of 100 per cent in Bangladesh) which would further limit the field of operation as far as product value is concerned.

Markets

Again we see a familiar pattern on the market side. A relatively low-value material is being converted through drying into a much higher-value product that can be sold into urban markets. One of the problems now facing the producers is the exact size of that market. This will affect decisions on future expansion or replication of the project.

There are two possible avenues that can be explored. The first is simply an expansion of the existing outlets. To this end MCC are seeking to deal with biscuit manufacturers in other towns such as Chittagong, suspecting that the Dhaka market is virtually saturated. The second option is to attempt to open up a different market for the coconut. The possibility that was initially identified was direct selling to urban middle-class consumers for use in cooking.

To achieve success in this market, the coconut will have to be promoted as a convenience food. This approach is unlikely to succeed for two reasons. Firstly, fresh coconut is widely and cheaply available in the urban markets. Secondly, those who could afford to pay for the dried product probably fall into a class that employs somebody else to cook for them and thus they have little vested interest in convenience foods.

The bulk export market is not accessible to the Surjosnato producers because the Bangladesh Government has banned the export of coconut. (They have made occasional sales to foreign organizations such as Traidcraft.) In any case, the Bangladeshi product could not compete on price with large-scale production from countries such as Sri Lanka.

There must be some doubt, therefore, about the opportunities for major expansion of the project or its replication elsewhere in Bangladesh because of limitations in the size of the market. We have seen the same constraints at work in Peru and are led to the conclusion that many of the suitable applications for the drier will not be directly replicable since they deal with high-value, low-volume products. It is in fact the technology itself which can be successfully replicated and it is inevitable that this will imply modifications to match individual circumstances. This indeed is the process that we have observed.

Profitability

ITDG's experience of rural agro-processing activities in Bangladesh is that the margins available are small. This is reflected in the coconut-drying project where the profit (after paying all fixed costs and depreciation) is around 5 per cent of the total sales revenue under the best possible circumstances. A substantial proportion of this profit (50 per cent in 1987) is used to pay off the loans, leaving the rest for distribution as dividends. In this particular project, the main benefit to the producers is in the form of wages and not dividends.

The loans from MCC have been interest free and without fixed terms. It has been largely left to the producers to decide how much to repay and how much to allocate to dividends. In a commercial environment the project would have to deal with fixed-term loans and high (16 per cent) rates of interest. This is, of course, assuming that it could gain access to such loans. However, after several years of development, the project yields an annual return of around 50 per cent of the capital invested in equipment, indicating that it would be viable even under commercial conditions.

The disincentives for the commercial entrepreneur are the decentralized organization of the enterprise and the high level of working capital tied up in it. Pre-purchase of coconuts for dry-season processing can require sums in the order of Tk 250 000 ($7150) which is equivalent to the total capital tied up in equipment. This type of project is, therefore, perhaps only likely to succeed in an environment such as that sponsored by MCC or a similar NGO.

2. PROSPECTS FOR THE 1990s

The processing facilities described in the case studies were established between 1980 and 1985. As ITDG is in the fortunate position of still having contacts with the collaborators involved, it is interesting to review the situation in the 1990s. Have the units survived? What problems have been encountered? Has the basic technology been replicated or disseminated to other countries? All these considerations are very important when trying to meet one of ITDG's objectives—sustainable development.

Perhaps the most important and startling fact is that all the plants reviewed continue in operation, in most cases with increased production.

The oldest of the units at Cuatro Pinos Co-op in Guatemala remains much as described. During a recent visit the idea of converting the ten drying chambers to a perhaps more appropriate tunnel drier was discussed. While a tunnel drier will result in some reduction in employment, this is seen as necessary because Cuatro Pinos is now finding shortage of labour a problem. While the drying facility remains as it was, there has been major growth in the co-op as a whole. Annual turnover is now measured in millions of dollars, but despite the fact that the technology originated in Guatemala there has been no further replication. This is somewhat difficult to understand. The equipment has proved reliable, cheap and economic. A local capacity has also been developed to build both cabinets and gas-fired heater/blowers. There is a good knowledge of the technology in the country and specialist advice of high quality is available from INCAP. Clearly, either no other opportunities exist for dehydration-based industry, which is hard to believe, or some essential trigger to replication is missing.

In Peru, Yerfil continues to operate two driers in Lima and Tarma. The owner plans to build a larger drier in Lima and with rapidly rising fuel costs (700 per cent increase in 1990) is becoming much more aware of efficiency. For this reason he is considering going back to the more costly, in capital terms, but more fuel-efficient semi-continuous chamber. He also attempted to build a Peruvian version of the Jetaire Zeta which he admits is far more efficient in terms of air delivery, fuel and electricity consumption than the Dolorien model. A chassis and heat exchanger assembly was constructed—the idea being to import from Bensons all other components. Unfortunately the imported kit became lost during a customs strike that lasted almost a year. A new Meisa packing machine has been purchased which not only produces teabags at the rate of 160 per minute but also puts them in their outer envelopes and counts them out in groups of 100. Yerfil now has a sophisticated printing facility on-site, producing all its own packaging.

From the development viewpoint it is interesting to note the changes that have taken place in Yerfil over the years. The whole plant is now much smarter, better ventilated, well painted and lit. In addition a small canteen providing lunches has been built and the workers are being assisted by the company in a self-build housing scheme.

The total workforce is now about 35. Little is known of the farmers growing herbs for the company except that most of those involved when the case study was carried out continue as suppliers, which seems to indicate the financial attractiveness of such crops and the security that growing with a contract provides.

Dolorien Engineering still builds heater/blower units and the owner estimates that about 50 have been sold. Several new technical innovations have been made. The company exhibited at the Peruvian International Industrial Fair in 1989 and received particular mention in the press for its driers. This is apparently resulting in many more orders.

In Peru, then, substantial dissemination has occurred; the technology is now firmly established as a viable alternative to imported driers.

Little is known of the situation in Colombia except that the Morenos plant remains in operation with increased sales. As far as we know, no replication has occurred and the drier remains the only one in the country. This could well be because the drier is 'locked up' in a commercial plant and the owner would clearly have little incentive to allow access to potential competition. He has, however, always been very helpful and open when ITDG has requested entry for people from other countries interested in the system.

The desiccated coconut operation in Bangladesh continues some ten years after its establishment and the mechanical drier is regarded as an essential part of the process. Recently it has been found that the somewhat rustic wood-fired heating system is not sufficiently reliable or controllable and with the increasing availability of reasonably priced electricity in rural areas consideration is being given to converting the tray drier to electric heating. ITDG has agreed to provide technical assistance if required. Again in Bangladesh, despite proving its usefulness, the tray drier remains the only one in operation. Very recently, however, another NGO has expressed interest in the system for other applications.

Testing the new drier in Cuba

Since the case study visits were carried out, further dissemination of the tray drier has taken place. One unit was constructed in Eire by the Shannon Development Authority which was interested in examining the feasibility of producing organically grown dried herbs. In order to carry out the feasibility study it was necessary to produce small amounts of dry product for price and quality evaluation. The low cost of the ITDG drier allowed the Authority to produce high-quality costed trade samples with the minimum investment. As might be expected, with the high labour costs in Europe, the drier was not considered suitable for use in the project when compared to more sophisticated automatic units. It is of interest to note, however, that it did allow the SDA to overcome a bottleneck in their study and that according to the trade reports the final products were of extremely high quality.

Tray driers have been established in several other countries. In Sri Lanka one drier has been built and successfully produced good quality, export-grade semi-crystallized fruits. It is now planned to carry out trials on drying medicinal plants.

In Cuba, the Cuban Federation of Women are using the system as part of a major medicinal plant programme which is supported by FAO. Three small (250 000 BTU) and one larger (400 000 BTU) Jetaire heaters have been delivered to the project. News of the planned development of the project is limited, but the Cuban technologists are carrying out some interesting work on the effects of cabinet insulation and recirculation of exhaust air on fuel consumption. A larger tunnel drier is also to be built.

In Ecuador, a co-operative is starting to examine the suitability of the drier to produce dry fruits in a very recently built unit. A drier has also recently gone to Lesotho where it has been used in a baseline study on the economic viability of asparagus drying.

The drying seminar in Peru

There has thus been considerable transfer of the technology with units based in nine countries. It seems that these small locally constructed driers meet a need in many developing countries. But only in Peru, with 50 units operating, have we seen real uptake and local adaptation of the technology.

What, then, makes Peru different? This is an extremely important question that needs to be addressed if, as is hoped, the technology is to provide wider benefits to poorer people. If a technology is to move forward and spread without relying upon continued meetings and collaboration with individuals, it needs to be marketed. In Peru, it seems that ITDG advertised the product as an alternative in drying technology by means of the seminar/workshop at the International Potato Centre in 1985 which was attended by some 40 representatives of NGOs, institutions, entrepreneurs and engineering workshops. They had, in Walter Rios, a 'technology salesman' who could advise on the use of the equipment and provide sound advice and in Dolorien Engineering a mechanism to provide and advertise the equipment.

ITDG is initiating a two-year strategy in the Americas to test the conclusions reached above. Three seminars are to be held in Central America, Latin America and the Caribbean to which representatives of NGOs, local institutions, development agencies, entrepreneurs and engineers will be invited. The seminars will be of a practical nature and present alternatives in small-scale industrial drying — not simply the ITDG type of tray drier. A range of dissemination material in both English and Spanish has been prepared, including a technical brief that can be sent as initial information to interested enquirers. To overcome reliance on imported heaters it is hoped that the Peruvians will allow transfer of the design of their heater, via drawings, to other countries. This area is at present (1990) under negotiation.

The first test of this dissemination strategy took place in Guatemala in 1990 at an ITDG-funded seminar attended by 35 representatives from seven countries. This resulted in Costa Rica and Panama holding similar in-country seminars and interest in setting up a prototype drier in Panama.

When ITDG's dissemination strategy has been further tested, it is hoped that we shall be in a better position to assist other countries which express an interest in establishing small industry based around the oldest food preservation method known to man — drying.

3. COSTS AND BENEFITS

Chapters 1 and 4 make it clear that only the production of relatively high-value products will be able to support investment in mechanized drying on the small scale.

The production of high-value commodities inevitably implies entry into the commercial marketplace, targeting either the better-off domestic consumers or industrial concerns. In general the level of investment required is likely to be beyond the immediate reach of a group of poor rural producers and they are unlikely to have the requisite skills in organization, accounting, engineering or marketing.

The drying process itself, on the scale discussed in this book, generates little direct employment; each drier can be operated by two to three people. What the technology can do is to link the producers of the raw materials to the end market, generating on-farm employment and income and a range of other benefits — employment in packing plants and in small workshops, by-product industries etc. Profits from the process accrue, of course, to the owners of the plant, whoever they may be.

For an organization like ITDG, with its focus on income and employment generation for the poorer sections of the community, the key to success is the dissemination of the technology so that the benefits reach as large a number of people as possible. A dilemma appears as to the best means to go about this process. In simplistic terms this boils down to the debate familiar to many development organizations — 'entrepreneur versus co-operative'. This chapter attempts to distil some of the experience from the case studies, showing the advantages and disadvantages of each dissemination pattern and some of the lessons that have been learnt.

It is difficult to make the distinction between who is an entrepreneur and who is not. Any person or group getting involved in the sorts of markets mentioned above has to behave in an entrepreneurial manner; any operation is going to have to show a profit. In this discussion we distinguish between concerns whose main objective is to benefit a group of poorer producers and those explicitly set up for profit. Thus the case studies from Bangladesh and Guatemala fall into the first category and those from Peru and Colombia into the second.

In the rest of this chapter we look at the distribution of benefits in the various cases, issues of ownership and control and possible patterns for further dissemination activities.

Benefits

We can look at benefits from the introduction of the drier under two general categories — employment generation and income generation. Employment can be generated directly on the farm, in the drying plant and in 'downstream' processing such as packaging. Income can be generated as farmer's profits and as profits to the plant (sometimes split into a drying component and a downstream component).

In Peru, the growing of camomile is extremely profitable in both percentage and absolute terms. A cost breakdown for one tongo (769 square metres) is given in Table 5 below.

Table 5. Tarma camomile production costs

Item	—— Cost ——		—— % ——	
Seed (1kg)	5.41		1.7	
Labour (23 days)	37.30		11.5	
Fertilizer (50kg)	4.05		1.3	
Pesticide	3.24		1.0	
Other charges	12.97	62.97	4.0	19.4
Margin		261.35		80.6
Selling price		324.32		100.0

(All costs in 1987 US$)

This corresponds to an annual production of 2.4 tonnes of camomile which is sufficient to supply the existing drying plant for ten days at full capacity. The latest word from Tarma suggests that the drier is working at maximum capacity, giving an annual return to the farmers of approximately $US8,000 at 1987 prices and assuming a 300-day working year for the drier. Perhaps a better indicator for the farmer is the return to labour which works out at $11.35 per day. This compares rather well with local wages of $1.71 per day!

Of course, the absolute level of returns is not the key issue. In an area like the Picoy valley, any new activity for the farmer must displace some other activity and it is the difference in returns that defines the true value of the new crop. From the degree of enthusiasm being shown for growing camomile, we can assume that the returns do compare favourably with other cash crops although it is very difficult to establish actual figures.

Despite these very favourable returns, the farmer's profits amount to less than 10 per cent of the total profit to be made from the manufacture of camomile teabags. The vast bulk of the profit (84 per cent) is actually made by the packaging process which is based in Lima. The other 6 per cent or so accrues to the owner of the drying plant in Tarma. To what extent is it possible to bias the returns more in favour of the farmer? The packaging process depends upon the supply of pre-printed boxes and labels (printed within the packing plant but using imported materials). It also employs other equipment such as teabag-making machines. Perhaps more importantly, the selling operation requires a presence in Lima where the main market exists.

The question of how much the farmers want to be involved must also be raised. Would they, for instance, want to run the drying plant or are there other uses for their time? Since the farmers seem unwilling even to travel into Tarma where they could sell at least some of their produce at a higher price, they presumably have other preoccupations.

An interesting parallel can be drawn from the Guatemalan operation. The Cuatro Pinos co-operative functions as an agribusiness in which the farmers are effectively shareholders. The farmers' direct participation begins and ends with the working of their own land. It appears that they are happy to sub-contract the processing and marketing sections of the business to professionals who also arrange for the labour required in the plant.

Organizationally, Cuatro Pinos had a much more difficult birth than the business in Peru. The case study and the introduction refer to the amount of support received in the early years, a situation that is not likely to be repeated very often. The advantages of operating in the entrepreneurial environment can be seen as efficient organization, detailed local market knowledge, single point of contact etc. The drawbacks include dividing the benefits and the potential for the entrepreneur to exploit the workers and producers.

In Peru, it could be argued that the presence of a profitable commercial organization based around the drier was a spur to other investors to get involved in the technology and thus to spread the benefits. ITDG originally imposed an open-door agreement on Yerfil, allowing any interested party to inspect the drying machinery. Yerfil have cheerfully co-operated over several years. The NGO is thus not totally powerless when it has something that the entrepreneur wants. Yerfil has, however, recently engaged in a significant amount of labour shedding. The owner presumably sees this as necessary in order to regularize his operation and to maintain a competitive price. Any organization selling into an open market will have to show certain entrepreneurial characteristics in order to remain competitive but the co-operative or producer group will not necessarily have profit maximization as its main objective.

Analysis of the Surjosnato project in Bangladesh shows that the level of profit distributed among the members is in the order of 5 per cent of their annual wage income. Whilst this is not insignificant, it is obvious that the major benefits are through employment generation.

For ITDG the main lesson that comes out of past experience is the need to tailor any inputs to particular circumstances. In Bangladesh we would be extremely cautious about getting involved with entrepreneurs. The country has such a huge pool of un- and under-employed labour that the owner could easily squeeze the workers in the drive to profit maximization. Indeed it is difficult to see any other easy way for the project to increase profits.

In Peru we would ideally be looking for an organized group of small farmers to develop their own drying and marketing operation; yet we can see that they might face considerable constraints in terms of access to credit, supplies of packaging materials and marketing skills. Perhaps the most important factor would be the willingness of the farmers to become involved in off-farm activities. Here the entrepreneurial route has benefited both farmer and plant owner but the plant workers have found themselves in a very vulnerable position.

It is most difficult to see the benefits to ITDG's target groups in the Morenos plant in Bogota. Wage employment is undoubtedly created but could disappear overnight if Sr Moreno turned his attentions to other projects. Yet the total input from ITDG was to post a set of drawings. In terms of returns to our own effort expended this may be the best intervention of all. This type of calculation is, however, fatally flawed because of the issues of ownership and control.

There is still a school of thought that says that since entrepreneurs are the most dynamic group of possible collaborators—because they get things done—working through them is the most effective way to disseminate new technologies. This may well be true but it is not enough. The trickle-down theory of development was largely discredited in the 1960s but that message does not seem to have trickled down to some people. The NGO dealing with the entrepreneur has to attempt to find some sort of leverage in order to secure a portion of the benefits for the community. Perhaps even more difficult is ensuring that those benefits persist over time.

Dissemination

The discussion of benefit distribution is largely an ideological one. In seeking to disseminate the technology, i.e. to spread the benefits more widely, it is also necessary to be pragmatic and to some extent to play the numbers game. It is really a question of balancing the benefits and disadvantages discussed above. It has to be borne in mind that, in most countries, a truly successful technological innovation will find its way into the entrepreneur's hands eventually. In a competitive market the entrepreneur will almost inevitably squeeze out the co-operative through labour cost reductions or other methods.

Organizations like ITDG can perhaps play entrepreneurs at their own game by extracting concessions in return for the desired information. These might include 'open door' agreements for a specified period of time, agreement to take part in seminars or workshops, or access to records so that data on the operation can be gathered. What is very difficult is to ensure that the entrepreneur's profit maximization strategy does not impact negatively on employees or suppliers.

Dissemination through the co-operative route takes far more time and effort unless the co-op is already well-established and with a degree of technical proficiency. It also requires marketing skills that may not be immediately available. The entrepreneur presumably has a market in mind from the start whereas the co-op is more likely to start with a product and then look for a market. Obviously the benefit flows look much better from an idealist's point of view but it is critical that the group concerned adopt a hard-nosed commercial attitude. This is particularly so when the market is competitive.

The range of support needed from the development agency is often much wider than purely technical skills. In fact, in our experience, these factors are often secondary to the commercial ones. The Guatemalan case study shows that these commercial skills can be bought in; the question arises as to what scale of operation is required to support these costs. (In Cuatro Pinos, the parsley drying plant accounts for a very small fraction of the co-operative's turnover.)

4. THE ECONOMICS OF SMALL-SCALE DRYING

Throughout the early stages of its Tray Drier Programme, ITDG had to answer considerable criticism of its focus on drying comparatively high-value crops for urban markets rather than the drying of staples. In this chapter Saleha Begum, formerly of the Institute of Development Studies at the University of Sussex, examines the economics of crop drying and concludes that opportunities, where they do exist, lie largely in adding value.

It is useful to distinguish between two types of drying: *product preservation*, where drying is required to prevent deterioration and therefore loss of value, and *product refinement*, where drying is a stage in processing a raw product into another form for final consumption. Usually, this 'product refinement' drying concerns preparation of a dehydrated product and in the food science and technology literature these two types of drying are very distinct subjects. However, both are relevant as potential sources of income generation for the rural poor.

In product preservation, no increase in the value of output over and above that gained by traditional methods can be achieved, except through loss prevention; therefore the value of losses in traditional drying dictates the maximum costs that can be incurred in improved drying methods.

In product refinement, value is added in drying and the drying operation is a necessary and integral part of a larger production process. Therefore, the costs of drying must be evaluated together with all the other inputs required to produce and market the product. Losses of quantity and quality of dried product will certainly help to determine which drying technique is adopted but the value derived from drying is not dependent solely on loss prevention. It is in product dehydration that improved small-scale drying techniques have most likelihood of being adopted.

There are several factors that contribute to this but the overriding technical factor is that dehydration is more energy-intensive because much more water generally has to be removed than with drying for preservation. Ambient conditions will often dictate the use of some form of drying other than open sun-drying for dehydration but this need is not so clear where preservation is concerned. In fact, in the case of product preservation, the introduction of small-scale driers is usually attempted as a substitute for open sun-drying. However, both of these sets of circumstances have been the focus of project research, design and development work.

Product preservation

Moisture-related deterioration of produce is often regarded as one of the most serious causes of post-harvest food losses. In the last decade, increasing attention has been given to the prevention of these losses as a main component of attempts to achieve food self-sufficiency in developing countries. The principal strategy has been the introduction of new drying techniques to replace traditional methods, most commonly open drying in the sun, which are alleged to result in high levels of food losses.

Traditional drying is often inadequate because the moisture content is not reduced to the optimal level for subsequent post-harvest operations — transport, storage, and processing. Losses occur in three ways:

- high moisture content increases the incidence of physical losses and qualitative deterioration because it encourages biological attack (particularly by insects and micro-organisms);
- chemical changes in the produce result in loss of dry matter and qualitative deterioration;
- physical losses occur because under- (and sometimes over-) drying are responsible for further processing being done at the wrong moisture content—for example, rice mills are usually designed to process paddy at 14 per cent moisture content and any substantial deviation from this level is likely to cause a reduction in the yield of whole, polished grains of rice.

The failure to dry produce properly using traditional methods is not usually ascribed to either neglect or ignorance but to inherent constraints associated with open sun-drying. Rain prevents drying; and cloudy conditions, with high relative humidities and reduced solar radiation, impair the effectiveness of open drying. Bad weather often results in irreversible qualitative deterioration. The rate and uniformity of drying are two further factors which affect quality of drying and which are difficult to control using traditional methods. Physical losses also occur on the drying floor, because of scattering and animals feeding on the produce. Physical losses occur after drying because of contamination and infestation affecting the unprotected grain during open drying.

Clearly, the significance of any of these constraints depends on local conditions at the time of harvest. However, in many situations increased yields and changed seasonality of production have raised the probability of food loss resulting from inadequate drying.

Because of the complexities in accurately measuring the various types of food loss potentially associated with poor drying, there is relatively little reliable evidence on the extent of such losses. Nevertheless, loss prevention has been the driving force behind many attempts to introduce small-scale drying techniques. The implicit assumption has been that the additional costs associated with improved drying techniques will be less than the benefits from the prevention of losses.

It has usually been recognized that physical losses on the drying floor are negligible and that it is qualitative deterioration — leading to discard, losses during subsequent processing and reduced unit price — which is the problem.

There are two basic reasons why field experience with the introduction of small-scale driers for product preservation has been less than encouraging. First, the level of losses with traditional drying practices has often been exaggerated. Estimating such losses accurately is difficult (see Russell 1980 and Greeley 1986) and worst case rather than average experience has influenced perceptions of the problem. In practice, this has meant that the benefits from improved small-scale driers have been overestimated.

Secondly, even when the level of losses in terms of physical indicators is known, the monetary value of benefits from loss prevention is often overestimated. In developing countries, the sensitivity of prices to quality characteristics can be relatively slight. The market will not bear a sufficient premium for a high-quality product to cover the additional drying costs. Moreover, even in regulated markets operated by public sector food-handling agencies where grading standards are defined, it is often not possible to implement the grading standards effectively. A further aspect of this valuation problem is that local perception of what constitutes 'quality' may differ markedly from international quality standards. For example, it has been observed that some south-Indian rice consumers pay a premium for paddy which has evidence of insect attack—because this is also evidence that the paddy is not freshly harvested but has been in storage for some time—and this is preferred because older paddy produces a better tasting rice, according to local palates.

In addition to these two basic reasons there are numerous other related considerations which have discouraged widespread adoption of small-scale driers for product preservation. Perhaps the most important of them is the seasonal variability in demand for drier services. Driers will be needed only when ambient conditions prevent sun-drying. The risk of investing in improved drying methods is that these conditions are not sufficiently common to show a profit on the investment. In the tropics, even short intermittent periods of sunshine will frequently be sufficient to prevent serious losses occurring. Some projects which have provided driers for commercial use have found that producers will hold back from the use (and expense) of the drier in the hope of a change in the weather. As a consequence either deterioration occurs before the produce reaches the drier, or drying capacity is insufficient when large quantities arrive at the same time. (This problem is compounded with communal driers if producers are unwilling to mix their produce.)

In some cases where inclement weather prevents sun-drying, producers have alternative strategies that can limit losses. In Bangladesh, for example, with paddy harvested during the rainy season, producers delay the harvest, change the method of stacking the harvested paddy, dry small quantities over the kitchen fire, and mix wet and dry paddy.

A second consideration relates to the fact that, for dietary staples in developing countries, the major part of production is usually for home consumption. In this circumstance no cash benefits will be derived from quality preservation so there is no financial motive for using, say, a communally owned small-scale drier. If complete discard of the damaged produce occurs there will be a real loss. But if, as is much more common, the effect of inadequate drying is simply to cause discolouration and off-odours then the cash-conscious producer-consumer will often prefer to accept this penalty rather than paying out for drying services.

A third consideration is that the monetary benefits from quality preservation through investments in drying relate mainly to consumer-observable quality characteristics which do not always fully reflect the overall product benefits from drying. For example, the prevention of infection by toxin-producing microflora may be a result of improved drying methods but this benefit will not generally result in a higher price for the output from the drier since it is not readily observable.

Similarly, with paddy, the fact that over-drying using uncontrolled traditional methods results in lower milling yields does not usually mean that selling prices for over-dried paddy will be lower since this internal damage to the kernel is not visible.

From a social perspective, it is worth noting that differences in price related to quality of food and nutritional worth of food are often not correlated; e.g. discolouration that lowers price does not necessarily lower nutritional quality. Likewise, higher prices paid for well-polished rice compared to undermilled rice in effect mean paying more for a less nutritious product.

A final consideration concerns the costs associated with operating small-scale driers. Often, such driers involve use of local raw materials, require space for permanent siting, and need labour time to operate them in excess of that associated with traditional drying. These costs are extremely difficult to estimate precisely because they are typically in non-market situations. There is a tendency for them to be given a very low value or even ignored by project staff even though they clearly represent costs to the producer.

This consideration is particularly valid for some solar driers which have thin drying beds and which are inherently less efficient in cloudy/rainy conditions and therefore require longer to dry produce. In practice, this has meant that, as far as drying for quality preservation is concerned, solar driers have not provided a satisfactory substitute for open sun-drying.

Collectively, these considerations impose severe limits on the feasibility of small-scale drying for quality preservation in many developing countries. They are all reasons either for the value of losses incurred in traditional sun-drying being less than it might at first appear or for the costs of preventing such losses being greater than is immediately apparent. The proponents of loss-preventing technical change through improved drying have, in general, been too strongly influenced by the technical requirements to preserve quality and not sufficiently well informed of the social and economic conditions that determine the costs and benefits of the proposed technical change. We can anticipate a more favourable climate for driers as the marketed surplus increases, as price sensitivity to quality increases, and as the opportunity costs of inputs associated with traditional drying methods increase. However, these changes are also likely to result in greater centralization of post-harvest operations with a commercial demand for large drying units as a part of large public or private sector processing, storage, and marketing operations—they are unlikely to offer substantial opportunities for small-scale drying that can be conducted by the rural poor.

Product refinement

Commonly this is where a dehydrated product is being produced, although there are some examples of drying other refined products—for example, drying paddy after parboiling. Attempts to introduce small-scale driers for product refinement have usually been as part of fruit, vegetable or fish drying projects. Often, this has been part of an income-generation programme for the rural poor.

In some cases, the drying project has involved the introduction of either new products or new marketing arrangements for existing products for which there is only a small market. Thus, while the drying unit may be a central part of the production process, the success of the overall project has depended on the development of a market. In this respect, such drying projects parallel handicraft projects in requiring a wider range of skills from project personnel, and ultimately (if successful) from project beneficiaries, to develop the market.

In other cases, the objective of projects in this area has been to improve the quality of dried products by avoiding the contamination associated with existing methods of traditional open sun-drying. However, even when the simplest solar drying methods have been used (e.g. by MCC in Bangladesh for dried coconut and fish) the produce has often not been price competitive in local rural markets because customers have not been willing to pay any premium for the better quality product from the improved drying practice. Consequently, these dried products have to be aimed at an urban market—large-scale caterers or middle-class consumers.

Thus, in both these cases the choice of drying techniques is only one, and perhaps not the most central, component of the drying project. Packaging and marketing arrangements will be most crucial in determining demand and it is these arrangements which have most frequently been problematic. Project success does not depend only upon the successful organization of rural skills, resources and labour, as it has to be complemented by urban trading skills. In practice, market development often means rural producers being dependent upon marketing agencies over which they exercise little influence.

Despite these problems with drying for product refinement, it is this application of drying technology which offers the best opportunities for ITDG programmes. While such opportunities are likely to be greater in less-poor developing countries (Peru more than Kenya or Sri Lanka, with Bangladesh the least promising) because of consumer demand patterns and the development of marketing infrastructure, some opportunities will exist in all countries. Development of effective projects in this area will depend upon successful and sustainable development of markets which are prepared to buy better products than traditional refinement methods can deliver. Technical assistance from marketing specialists will therefore be just as important as food technologists' services.

References
de Padua, 1976, 'Rice Post-Production Handling and Processing: Its Significance to Agricultural Development'. Paper presented at the International Workshop on Accelerating Agricultural Development, SEARCA, Laguna.

Russell, Donald G., 1980, 'Socio-economic Evaluation of Grains Post-production Loss-reducing Systems in South-east Asia.' Paper presented at the E.C. Stakman Commemorative Symposium: 'Assessment of Losses which Constrain Production and Crop Improvement in Agriculture and Forestry', University of Minnesota, Minneapolis.

Ryland, G.J., 1985, 'The Economics of Grain Drying in the Humid Tropics'. Australian Centre for International Agricultural Research, Canberra.

Greeley, M., 1986, 'Rice in Bangladesh: Post-harvest Losses, Technology and Employment'. D.Phil Thesis, University of Sussex.

5. TECHNOLOGY CHOICE

This fairly quick analysis of the range of technical options available to those interested in setting up drying operations indicates why small low-cost driers that can largely be built in-country meet a market need.

The choice of a particular drying system is not simply a matter of throughput and cost. Other factors that may need consideration include:

- local climate: possibility of solar drying, humidity;
- value added per unit weight dried;
- sensitivity of product: flavour, colour, case hardening (see Appendix);
- fuel availability and costs;
- final quality demanded in market;
- labour costs;
- working patterns; social acceptability of shift work;
- packaging: is the product one that will absorb moisture rapidly?

If we exclude specialized and costly systems such as spray, freeze, foam mat and roller driers as being beyond the scope of this book, a range of technical options appropriate to small industry in many developing countries becomes clear. These are:

1 simple solar driers, either cabinet or with a collector;
2 mixed solar/biomass-fired driers;
3 wood or other biomass driers;
4 small batch cabinet driers, fuelled and with forced air;
5 small semi-continuous tray driers;
6 tunnel driers;
7 moving belt driers.

All of the above are applicable to the dehydration of solid material in fairly thin beds, usually on a tray.

1 The simplest option is the type of drier being used by the women described in the Bangladesh case study to produce desiccated coconut. This is often called a Brace drier and is illustrated together with a slightly more complex design that has an additional solar collector which improves

Brace-type solar drier

efficiency. The external collector driers can also be used, if connected to a wooden cabinet, to keep the product out of direct sunlight. This can be very important in the case of a material whose colour is affected by light.

All solar driers suffer from the obvious disadvantage of not operating at night or in dull weather. In addition, the following problems are encountered:

- in very humid climates the temperature rise inside the cabinet is insufficient to dry the food completely;
- drying times are usually long which means that moulds and yeasts have plenty of time to grow, particularly if the food is not dry at the end of the day and has to be left overnight;
- unless a group of people work together, the capacity is often too small to prove viable.

All these negative aspects do not mean that small solar driers cannot have applications, although the number of recorded viable, sustainable projects using solar driers known to the authors is very limited. The Bangladesh case study is one of these. Another example comes from Honduras where a local NGO *Pueblo a Pueblo* has assisted in establishing a dried cashew fruit unit that is now exporting about five tonnes each year.

If the climate is good and a reasonably high added value product is involved then a group running a number of small solar driers can be viable.

2 A slightly more complex alternative, which aims to overcome some of the disadvantages of a simple solar drier, is a mixed solar/ wood-fired unit, often called a McDowell drier, after the originator. Here, a tray of material is covered with a roof that allows solar drying when there is

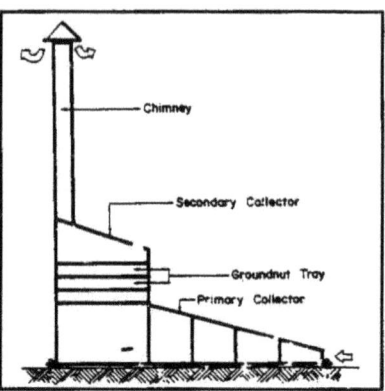

External collector drier

sufficient sunshine. If the weather deteriorates or the product is not completely dry by nightfall a fire is lit in the firebox. The heat is then transferred to the drier by means of metal heat-exchange tubes running under the tray. Clearly, running costs for a McDowell drier will be higher than for a straightforward solar system as fuel is involved and the initial capital cost is greater. Profitability will thus depend on adding sufficient value to cover these costs, possibly by having a product of higher quality than would be possible using the solar drier.

3 A number of designs have been proposed, tested and used for wood or biomass driers which, like the McDowell unit, have some sort of heat exchanger transferring the heat from the fire to the product.

None of these can, in general, be considered appropriate for small industry owing to their disadvantages of low output, lack of control and limited air flow. Their use is more applicable to groups in rural areas wishing to gain additional income from preserving small quantities of produce for sale or to reduce the loss of their higher value crops such as spices. As we move into the more informal small industry sector, larger, more efficient, controllable driers become essential. These are usually gas- or diesel-fired and have forced ventilation.

4 The simplest driers of this type are batch driers in which heated air is passed into a cabinet loaded with trays until the whole charge is dry. Various degrees of sophistication are available including variable air recirculation controlled by humidity and heat recovery from exhaust air.

McDowell drier

The ITDG batch drier being used in Peru is perhaps the simplest and cheapest system possible, consisting of a wooden box filled with trays and fed with hot air from a diesel heater/blower. Fuel efficiency and air-flow patterns are, however, far from perfect. Larger commercial batch tray driers are readily available that are highly efficient in terms of fuel efficiency, drying rates and personnel requirements. Some consist of two drying chambers at different temperatures, trays being moved semi-automatically from one to another to suit the product's particular drying needs.

5 As throughputs increase, semi-continuous driers begin to be more commonly seen, with the ITDG semi-continuous unit being probably the smallest available.

6 At a larger scale there are tunnel driers. Here carriages loaded with trays pass down a tunnel, normally against the flow of hot air. Again, various levels of sophistication exist in terms of control and handling.

7 At the top end of the scale continuous band driers are used and in developed countries have all but replaced tunnels. Here the product is fed continuously on to a moving band that moves slowly through the drier to discharge dry material at the exit. Such units are very fuel efficient, have

Table 6. Drying technologies compared

Type	Cost (US$)	Capacity wet kg/24-hour	Investment per kg dry capacity (US$)	Fuel efficiency	Labour level
Brace solar	50	10	50	n/a	v.low
Solar cabinet	70	30	23	n/a	v.low
McDowell	170	40	43	v.poor	low
Wood-burning	340	80	43	v.poor	low
ITDG batch	3 400	240	140	poor	high
ITDG semi-continuous	6 800	360	190	medium	v.high
Commercial tray cabinet (small)	85 000	500	1 700	good	high
Commercial tray cabinet (large)	170 000	2 500	680	good	medium
12-carriage tunnel	145 000	6 000	240	good	low
Moving band	800 000	48 000	170	v.good	v.low

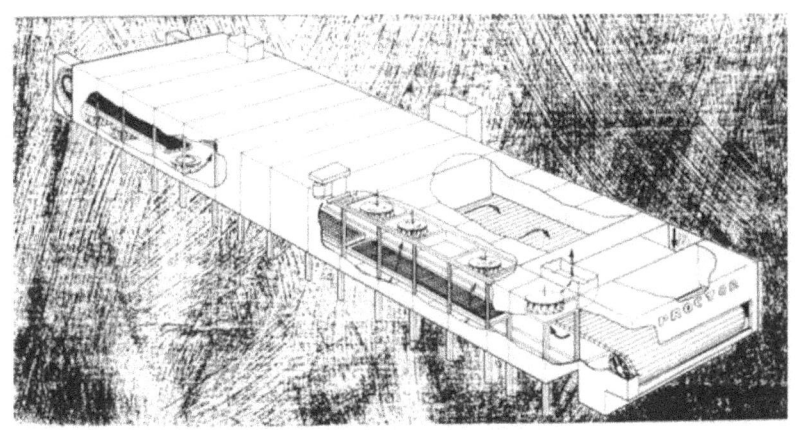

A typical conveyor drier

complex systems of control, require a carefully managed system of raw material supply and need little labour.

In Table 6 these options are compared in terms of approximate cost, capacity, capital investment necessary per kilogram of dry product, fuel efficiency and amount of labour required. A clear distinction can be drawn from this very approximate data.

- The role of solar, mixed and wood-fired driers seems to lie with small rural groups needing to dry and refine small quantities of foods. Capital costs are low—perhaps the most important factor for such groups where credit may be difficult or impossible to obtain. While labour is needed, it can be supplied within the family (keeping a fire going, for example).

- The low cost, in comparative terms, of the ITDG semi-continuous tray driers, coupled with the fact that they can be built in-country — so reducing reliance on external sources for spares — may account for much of the interest they are causing. However, such driers are not highly efficient and need a lot of labour. These factors would be of more importance in Europe than capital cost, the reverse being true for an entrepreneur in a typical developing country.

APPENDIX: DRYING THEORY

A basic knowledge of drying theory is essential for selecting drying systems and for troubleshooting on existing equipment. It is, however, our aim in this appendix to introduce as little technical jargon as possible and to present information that has been found useful in our own work in a form that is easy to understand.

Definitions
Moisture content
The moisture content (MC) of a material can be defined in two ways — on a wet or dry basis. Throughout this book we have used wet basis moisture contents.

The wet weight basis (WWB) moisture content is defined as the weight of water in the sample divided by the total weight of the sample (water plus dry material).

Humidity
The amount of water in the air is described in terms of humidity rather than moisture content by weight. Throughout this book the terms humidity and relative humidity are used interchangeably although this does not follow the strict scientific definitions given below.

The actual amount of water vapour in the air at any time is known as the *absolute humidity*, which is usually expressed in kilograms of water per kilogram of air (or pounds per pound). This measure is not used much in this book.

The *relative humidity* (RH) of the air is the actual humidity divided by the maximum amount of water that the air can hold (the humidity at saturation). This is always expressed as a percentage and depends on the temperature of the air as well as the amount of moisture in it.

When air is heated, the absolute humidity remains constant because the heating does not add water to the air or take it away. The relative humidity falls because the hot air is potentially able to absorb more moisture — it contains the same amount of water but is less saturated.

Drying rates
There are usually two separate phases in the drying of a wet product. In rather simplified terms, the first phase corresponds to the removal of

surface moisture. If the temperature and humidity of the air meeting the product are constant then water is removed at a constant rate. This phase is logically known as the *constant rate* period.

As the drying process proceeds, water must be removed from inside the product. This gets progressively more difficult as the moisture has to move further and further to reach the surface; the rate of drying gradually decreases. Thus we have the *falling rate* period.

It is worth briefly mentioning here the phenomenon of *case hardening*. This is the term used when the outside of the product dries relatively rapidly and forms a barrier to further movement of moisture from the interior to the surface. This will drastically reduce the rate of drying and may lead to eventual spoilage since the inside of the product remains wet. Some materials, potato for example, are very susceptible to case hardening; others are not.

It is not possible to dry materials to a zero moisture content but this is not the aim of the exercise. For most materials, moisture contents between 5 and 10 per cent will be low enough to ensure stability of the product against rapid rehydration or microbiological attack. After this point is reached, further drying is a waste of time, energy and money.

Practical drying
Batch and continuous driers
The simplest form of drier is the batch drier. A quantity of wet product is loaded into the drier and air is passed over it until the entire batch is dry. At this point the entire batch is removed.

This type of drying has advantages and disadvantages. It requires the minimum amount of attendance from the plant operators but it is not energy efficient and generally leads to lower throughputs. Figures 1 and 2 show (for a batch tray drier containing only three trays) simplified graphs of how the outlet air condition and the product moisture contents change over time.

When the product on tray 1, nearest to the air inlet, has become completely dry, the entire batch cannot be unloaded because the other two trays still require further drying. This is where the batch process becomes inefficient. Ideally one would decrease the supply of air as drying progressed thus saving energy. In practice, fans rarely have variable speeds and it is difficult to match the air supply to the condition in the cabinet.

Continuous drying is an attempt to overcome some of these problems. In a continuous drier, any product that has reached the desired

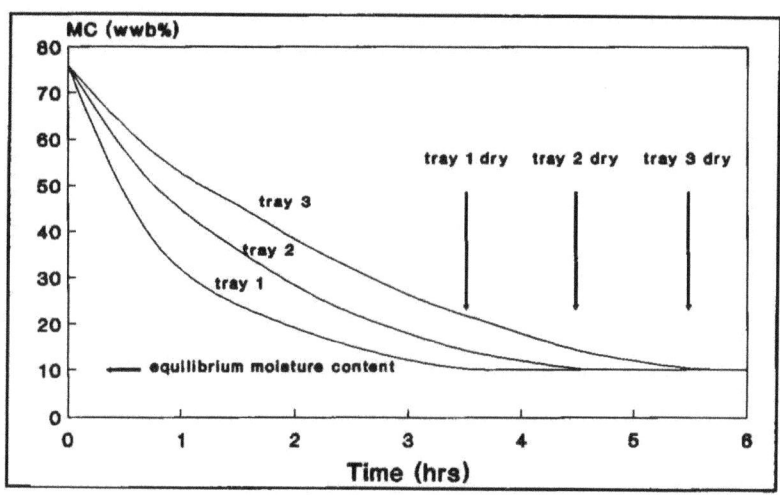

Fig. 1 Batch-drying: tray moisture contents

moisture content is removed and replaced with wet product. In a tray drier such as the ITDG design, the bottom tray is removed and a new tray of wet product added at the top of the cabinet. This is an attempt to provide the best drying conditions for each tray of product, to conserve energy and to avoid over-drying.

In a truly continuous drier the conditions at any point in the cabinet remain constant over time. We would expect to see a constant outlet air temperature. (Of course a tray drier is not truly continuous because trays are only changed every few minutes. Some variations will occur.)

Semi-batch drying

The distinction between batch and continuous drying has been spelt out but in fact pure batch or truly continuous drying are rarely encountered. In practice a form of 'semi-batch' drying is often employed. After some time in the drier, a portion of the product is withdrawn fully dried and replaced with wet product.

The Guatemalan operation is a good example of semi-batch drying with approximately half the trays being changed every three hours. The operators' schedule can thus be arranged to fit in with other activities and the amount of over-drying and thus energy wastage is reduced.

It can be seen from the case studies in this book that the camomile driers in Peru are close to being batch operations, the Guatemalan driers

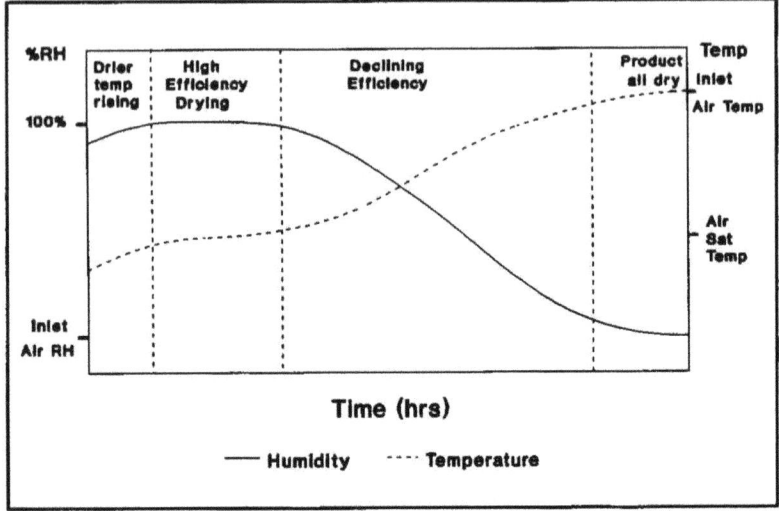

Fig. 2 Batch drying: air temperature and humidity

fall somewhere in the middle and the Morenos plant in Bogota approaches closest to truly continuous drying.

Drying curves

Drying curves can indicate a minimum drying time at a given temperature for a variety of materials but are rarely directly applicable to practical driers. (The bottom tray of a batch drier should follow approximately the relevant drying curve.) They can, however, provide useful information on the relative lengths of the constant and falling rate periods — the minimum drying time is an important constraint on drier capacity. (See below.)

Constraints

Various constraints can limit in advance the capacity of a drier. Perhaps the most common of these is the maximum air temperature allowed within the drier. This might be specified to avoid thermal damage to the product or to reduce the risk of case hardening.

The minimum drying time at the air inlet temperature, as given by the appropriate drying curve, places a lower limit on the residence time of the product in the drier.

A physical constraint such as the rate at which trays can be loaded or unloaded may also exist.

58

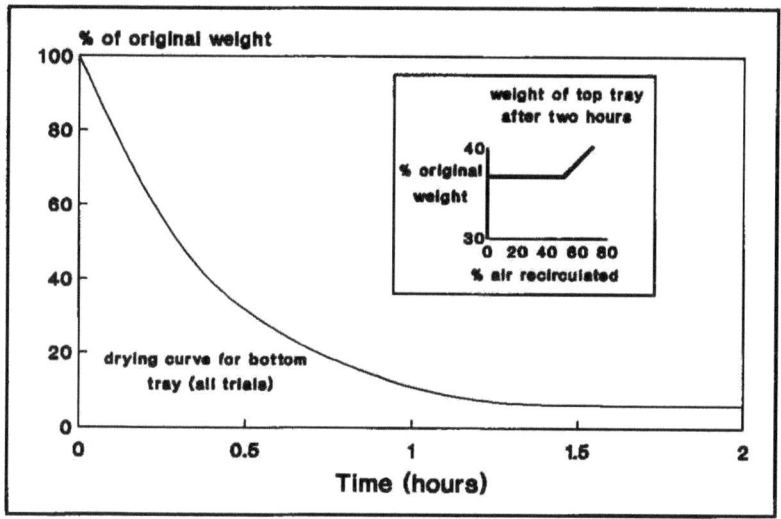

Fig. 3 Effects of air recirculation

Capacity

Capacity can be accurately determined only by experimentation. There are, however, some simple calculations that can be used to make an estimate. The most straightforward of these makes use of the so-called 'pickup factor' which is effectively the same as the drying efficiency.

The calculations should be performed for the worst expected ambient air conditions (highest temperature and humidity) to give a conservative estimate of capacity.

From this point on, the decisions become operational and economic rather than theoretical. A larger batch with a longer drying time implies a bigger, more expensive drier but less frequent attendance. It also implies more over-drying of the product closest to the air inlet. The converse is of course true for a smaller batch.

Air flow

The speed of the air through the cabinet is important because, in order to pick up the maximum amount of water, the air needs to mix well with the product. The literature on drying suggests that air velocities of 0.5 to 1.5 metres per second inside the drier provide the best conditions. Very high velocities lead to large pressure drops and the lifting of product off the trays; low velocities give poor mixing.

Tray loading

Trays in tray driers need to be loaded so that air flow around and through the product is as free as possible. Overloading leads to uneven drying as the air is channelled through a few free spaces. Underloading means poor utilization of the drier's capacity.

Air recycle

Except in the early stages of batch drying, the air leaving the drier is unlikely to be completely saturated with water vapour. There is often something to be gained by returning this air to the drier for a second pass over the product. This is known as recycling the air.

Recycling is a way of saving energy. The air leaving the drier is usually at a temperature above the ambient and therefore requires less heating to return it to the specified inlet temperature. Obviously only a portion of the air can be recycled or the relative humidity would rise to 100 per cent and no further water would be removed from the product.

Air recycle will not increase the capacity of the drier; in fact, after a certain point it will decrease it. It is purely and simply an energy conservation measure.

Drier performance

We can summarize the above sections by discussing the three parameters that most affect the performance of a drier, as follows.

Air flowrate

Low air flows give reduced capacities because the rate at which water is being removed is less than the possible rate of evaporation and the mixing between air and product is poor.

In most applications the air flow is determined by the use of a fixed-speed fan. Our evaluations have shown that the delivery of this fan is often the single most important constraint on drier capacity. The temptation simply to enlarge the drying cabinet to meet increased demand should be resisted. This is not to say that it will never be the solution but simply that a little further investigation is often needed.

Experience with the ITDG drier has shown that air deliveries of between 600 and 1000 scfm (cubic feet per minute at 20°C) or 0.3 to 0.5 cubic metres per second, depending on the product, give the best results. The quoted delivery of a fan is not always a good guide to the actual delivery since the pressure drops within a drier are rather unpredictable.

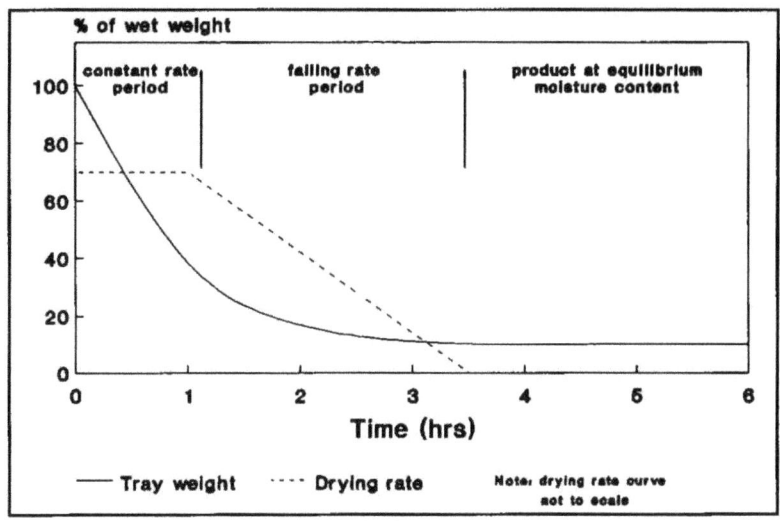

Fig. 4 Idealized drying curves

The best guide is to attempt to stay within the air speed limits quoted above. This is complicated by the fact that the actual path of the air through the drier is difficult to determine—is it flowing across the trays or through them? Rough calculations can be made and will usually prove sufficient.

Many of the large batch driers in Peru are suffering from a shortage of air. This has lead to increased drying times but only presents a problem when capacity is limited.

Air temperature
The need for high air temperatures to speed the drying process is obvious and the dangers of product damage and case hardening have been mentioned. A compromise must be struck and factors such as fuel costs inevitably need to be considered.

The Guatemalan experience illustrates a case where high temperatures are actually beneficial in terms of product quality. The enzyme which breaks down chlorophyll is destroyed leaving the dried parsley with a vivid green colour.

The MCC coconut drier in Bangladesh and the potato driers in Peru are good examples of cases where the maximum air temperature must be very carefully controlled to avoid damage and case hardening respectively.

61

Product residence time
In practice the final product quality is likely to be determined by look and feel rather than by accurate weighing. Whatever the method of determination, the product quality that is acceptable corresponds to a certain moisture content and the most efficient operation of the drier corresponds to a residence time that is sufficient to reach this level of moisture without leaving the product in the drier longer than is necessary. This is where batch driers become inefficient. When the last of the product is dry, the rest of it will have been over-dried to varying degrees.

External factors
There are external factors over which the user has no control which can be more important than operational practice. Initial investigations of the MCC drier in Bangladesh indicated that its capacity was dependent upon the tray loading. When the data was re-examined it was found that the ambient humidity was in fact the controlling factor and that tray loading had virtually no effect.

Optimizing performance
Using the maximum allowable air temperature, the size of the batch can be varied and average capacity at different loadings (in kg per day or per week) can be calculated. The overall capacity of the drier may or may not be affected by the different batch sizes. Loading and unloading schedules and the availability of labour need to be considered here and it may be worth investigating some form of semi-batch operation.

If the drier is fitted with a recycle loop, this should be used as soon as the outlet air saturation drops below 100 per cent. Figure 3 shows the result of some straw-drying trials performed by ITDG in a fifteen-tray cabinet. It can be seen that 50 per cent recirculation did not affect the drying time of the top tray in the drier but that 70 per cent recirculation did. With 50 per cent recirculation, 10 per cent fuel savings were achieved and with 70 per cent recirculation, 13 per cent fuel savings at the cost of slightly reduced capacity. In practice the degree of recycle can be gradually increased until the capacity of the drier begins to suffer.

Heaters and blowers
Energy is supplied to the drier installation for two purposes: moving the air through the cabinet and heating the air. In all the cases in this book, the

fan has been electrically powered. Heat has been variously provided by gas, diesel, kerosene and firewood. Electric heating is also possible.

Two methods of heating by combustion exist—direct and indirect. In a direct heater such as that in Guatemala, the fuel is burnt in the same air stream that eventually passes through the drier. The problem with this method is that any contaminants or unburnt parts of the fuel pass over the product and may taint it. This method is only really suitable for clean-burning fuels such as gas and, in the cases of oily materials such as coconut and fish, great care must be exercised.

Indirect heaters burn the fuel in a separate air stream from that passing into the drier cabinet. The Benson heaters and the units fabricated in Peru are of this type. The hot exhaust gases pass through a heat exchanger where they heat the incoming air. This type of heating is less efficient than direct firing but ensures a clean air supply to the cabinet.

The system adopted in Bangladesh, where the burning wood heats water which then heats the air, is yet more indirect and even less efficient. The main reason for adopting this system was to maintain a steady air temperature into the drier. Direct heating of the air by the wood fire would probably have implied large temperature fluctuations. The water in the system effectively damps out the variations in the furnace output.

Insufficient air supply was a major constraint on the capacity of many of the driers in the field. Sizing a fan for a particular cabinet is not easy; the delivery is determined by the pressure drop through the heater and the cabinet, which is almost impossible to predict. A few simple rules can be suggested.

- Avoid changes in the sizes of air ducts as much as possible. Generally, bigger ducts are better; a common fault has been to make the drier outlet much too small. Dampers can always be added cheaply to restrict air flow, but enlarging ducts is expensive.

- All ducts should be as straight as possible outside the drier. Inside the drier the air flow should be as turbulent as possible. Bends in ducts should be gentle, not sharp.

- It is much better to oversize the fan and then add a damper than to have to replace an undersized fan at much greater cost.